Defensive Information Warfare

Defensive Information Warfare

David S. Alberts

THE CENTER FOR ADVANCED
ACT
CONCEPTS AND TECHNOLOGY

Directorate of Advanced Concepts, Technologies, and Information Strategies

Institute for National Strategic Studies

National Defense University
Washington, D.C.

NATIONAL DEFENSE UNIVERSITY
- *President:* Lieutenant General Ervin J. Rokke, USAF
- *Vice President:* Ambassador William G. Walker

INSTITUTE FOR NATIONAL STRATEGIC STUDIES
- *Director:* Dr. Hans A. Binnendijk

DIRECTORATE OF ADVANCED CONCEPTS, TECHNOLOGIES, AND INFORMATION STRATEGIES (ACTIS)
- *Director:* Dr. David S. Alberts

CENTER FOR ADVANCED CONCEPTS AND TECHNOLOGY
- *Director:* Mr. Larry Wentz
- Fort Lesley J. McNair, Washington, DC 20319-6000
- Phone: (202) 685-3837 ◆ Facsimile: (202) 685-3664

Library of Congress Cataloging-in-Publication Data

Alberts, David S. (David Stephen), 1942-
Defensive Information Warfare / David S. Alberts.
p. cm.
1. Information warfare. 2. Defensive (Military science)
I. National Defense University. Directorate of Advanced Concepts, Technologies, and Information Strategies. II. Title.
U163.A395 1996
355.3'43—dc20 96-29321
CIP

First printing, August 1996

For sale by the U.S. Government Printing Office
Superintendent of Documents, Mail Stop: SSOP
Washington, DC 20402-9328 Phone: (202) 512-1800

Table of Contents

List of Figures

Introduction

Ervin J. Rokke

The Information Age carries implications for virtually all human endeavors, including the military profession. It's likely that these implications have or will produce revolutionary changes in warfare, but that issue remains unresolved among academics and military specialists alike. The search for answers, however, has generated a new intellectual excitement about military theory. It also has uncovered some preliminary notions about national security that require attention now.

In this treatise on defensive information warfare, Dr. David Alberts reviews one immediate, if narrowly focused challenge. The threat of information attacks, that is, "attacks on decision makers, the information and information-based processes they rely on, and their means of communicating their decisions," currently exists. With actual and potential practitioners covering a broad spectrum of sophistication and resources, it's a phenomenon which cannot be denied. In a very real sense, a new answer has emerged to a fundamental question in international politics: "What are the capabilities of the players?"

Dr. Alberts sets forth this new capability in a crisp, convincing manner. He relates it to "interaction arenas" ranging from military through economic and political to social and ideological; he describes its relevance for peaceful as well as conflictual relationships; and he notes its utility for all categories of "actors," both within and across national boundaries. It's a tool capable of creating dramatic results totally out of proportion with the inputs. Against this background of a complex, if not organic, capability, Dr. Alberts's prescription for policy draws interesting parallels with societal efforts to combat disease, drugs, and crime. Indeed, his defense resembles the human immune system to the extent that it involves a defense-in-depth strategy and works to "heal" the damage caused by information attacks as well as to prevent or blunt them.

Unfortunately, the technical precision which characterizes information warfare techniques is insufficient for answering two other fundamental questions in international politics, to wit: Who are the players? and What are their intentions regarding one another? While it is clear that information warfare techniques are available to empower a far broader spectrum of both nation and non-nation state actors, the extent to which this has occurred remains ambiguous. We simply don't know with precision who the information warfare players are or will be. In like manner, it is not yet clear how enthusiastic the new players will be about using their new-found weapon.

Accordingly, one hears that appropriate attention to information warfare defense may well have to await a so-called "information Pearl Harbor." Absent such an unfortunate event, Dr. Alberts acknowledges uncertainty about the willingness of the United States as well as other traditional actors to buy into information warfare defense. Publics and parliaments have grown accustomed to clear-cut opponents with measurable force structures. The threat described by Dr. Alberts is non-linear; it falls outside the traditional framework for guns versus butter calculations. His treatise does, however, provide a timely warning and a useful road map for meeting the major new security challenge of the Information Age. Hopefully, we will not respond too late.

National Defense University
August 1996

Preface

This overview of defensive information warfare (IW-D) is the result of an effort, undertaken at the request of the Deputy Secretary of Defense, to provide background material to participants in a series of interagency meetings to explore the nature of the problem and to identify areas of potential collaboration. This material, in briefing form, was provided to key decision makers in the Department of Defense and other agencies. In the course of these briefings, as well as other presentations to interested parties within government and in the private sector, much useful discussion was stimulated. Many suggested that this material needed to reach a wider audience to help achieve the increased awareness of this problem that will be necessary if we, as a society, are to come to grips with the challenges inherent in defending ourselves against these kinds of attacks. It is hoped that ACT's publication of this compact treatment of the subject will increase the attention that this subject receives and move along the public policy debate that is essential to progress.

David S. Alberts
Washington, D.C.

Defensive Information Warfare

Information warfare (IW) has become virtually synonymous with the revolution in information technologies and its potential to transform military strategies and capabilities. There is a growing consensus that national prosperity, if not survival, depends on our ability to effectively leverage information technology. Without being able to defend vital information, information processes, and information systems, such a strategy is doomed to failure.

Information Warfare and Information Strategy

Information warfare is often thought of as being defined by a particular target set—decision makers, information, information processes, and information systems. The "battlespace" associated with IW has been a constantly expanding one, moving far beyond traditional military situations. In some quarters, IW has even been associated with the leveraging of information technologies to achieve greater effectiveness and efficiency. This has stretched the meaning of IW to the breaking point and has sowed more confusion than enlightenment. For this reason, this treatment of the subject uses the term "information strategies" to refer to the recognition and utilization of information and information technologies as an instrument of national power that can be independent of, or complementary to, military presence and operations.

The scope, or battlespace, of information warfare and strategy (IWS) can be defined by the players and three dimensions—the nature of their interactions, the level of their interactions, and the arena of their interactions (see Figure 1, Scope of IWS). Nation states or combinations of nation states are not the only players. Non-state actors (including political, ethnic, and religious groups; organized crime; international and transnational organizations; and even individuals empowered by information technology) are able to engage in information attacks and to develop information strategies to achieve their desired ends.

The term "war" has been used so loosely in recent times (e.g., War on Poverty, War on Drugs, War on

***Figure 1.* Scope of IWS**

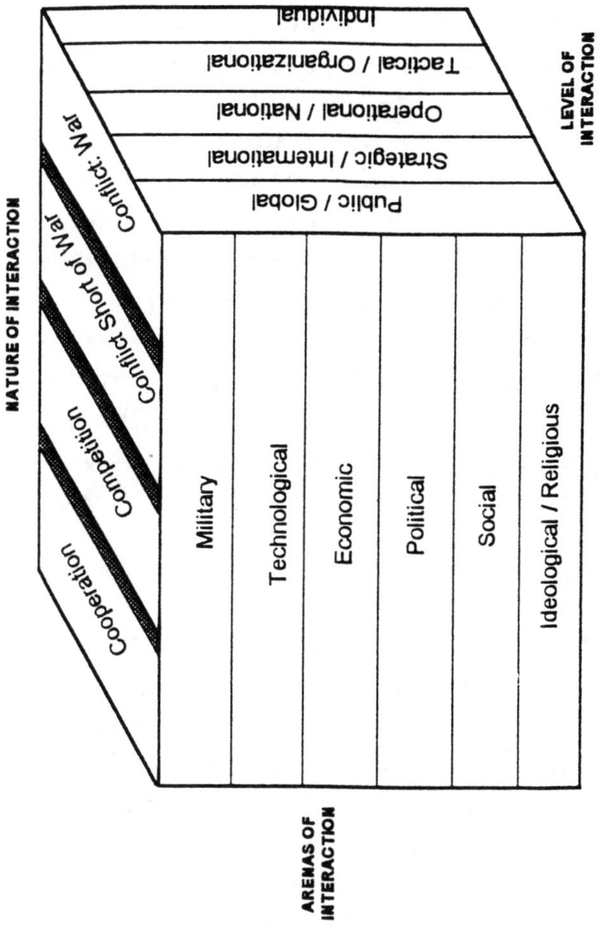

Crime) that is should be no surprise that IW has evolved over the past several years to become a "catch-all" term that encompasses many disparate activities, some of which have long been associated with competition, conflict, and warfare, and others that are of more recent origin. These include activities that range from propaganda campaigns (including Media War), to attacks (both physical and non-physical) against commanders, their information sources, and the means of communicating with their forces. Under this rather large umbrella that has become known as IW, one can find activities long associated with military concepts and operations, including deception, command and control warfare (C2W), and psychological operations (Psyops). Technological advances have added new forms such as electronic warfare (EW) and "hacker warfare."

The term "defensive information warfare" (IW-D) is used here to refer to all actions taken to defend against information attacks, that is, attacks on decision makers, the information and information-based processes they rely on, and their means of communicating their decisions. Strictly speaking, since these attacks can be launched during peace time at nonmilitary targets by nonmilitary groups, both foreign and domestic, the term IW-D should be IWS-D. However, IW-D is currently in wide use.

This overview of IW-D does not attempt to deal with the problems of defending against all of the different kinds of information attacks, but rather focuses its attention on the subset of IW that involves attacks against our information infrastructure, including what has become known as "hacker warfare" and in its more serious form, "digital warfare."

Analogies and Realities

Defending against information attacks has a number of characteristics in common with societal efforts to combat disease, drugs, and crime. Noting these similarities helps to put this problem into perspective, provides some potential useful lessons learned, and serves as a benchmark.

Before reviewing the specific similarities between combating IWS and these long-standing problems, it should be noted that, while eradicating information attacks may not be a realistic expectation, significant progress can be made in defending against all forms of information attacks, enough so that the risks can be kept at acceptable levels. Defending, as it is used here, includes preventing attacks, blunting attacks, and controlling the damage caused by attacks.

The problem of IW-D is similar to the problems encountered in the "wars" on disease, drugs, and crime in a number of dimensions. First, the solution to any of these problems requires the efforts of a number of organizations, both public and private. Second, it is unlikely, given the competition for resources, that any of these efforts will be fully funded. Therefore, we can expect that there will never be what those who have IW-D responsibilities think are a sufficient level of funding for IW-D programs. Third, these are not static problems. Drug cartels and criminals certainly learn from their mistakes. Even viruses "learn." Thus, defense forces will be continuously locked in a battle to keep up with attackers. Fourth, public awareness and concern will reach peaks, often accompanied by frenzied efforts to solve the problem. These relatively short periods of interest will be followed by longer periods when the urgency to solve the problem will give way to apathy. Maintaining funding and progress during these periods of waning public interest will be one of the key challenges of leadership in this area. Fifth, organizations and individuals will learn to make adjustments in their behavior to deal with IW attacks and their often unintended consequences. These adjustments will be made so that those organizations and individuals can accommodate some level of pain—a dynamic equilibrium of sorts—as the cost of doing business in the Information Age. Finally, solutions will, of necessity, involve compromises.

This is due to the natural tensions that exist among the various stakeholders. Tensions between the law enforcement and the protection of civil liberties are classic examples that have already arisen in the information domain.

Current Situation

Attacks on information systems are already a fact of life in the Information Age. Although a small portion of these attacks result in significant loss or damage, the vast majority of them result in little or no damage—the crime equivalents of trespassing, public nuisance, minor vandalism, and petty theft. It has been estimated that more than 90 percent of these attacks are perpetrated using available tools and techniques (based upon incidents reported to CERT), that only 1 attack in 20 is noticed by the victim, and that only 1 in 20 gets reported (these last two statistics were a result of a Defense Information Systems Agency (DISA) study and similar rates have been reported by others). However, it appears that reporting rates may be on the increase.

Of more concern is the presence of a technically feasible "strategic" threat. That is, the means exist to cause significant damage and disruption to U.S. public and private information assets, processes, and systems, and to compromise the integrity of vital information. Analysts also have no difficulty identifying groups with the motivations and opportunities to launch such attacks. Given our present vulnerabilities as a nation, a well-planned, coordinated IW attack could have strategic consequences. Such an attack, or the threat of such an attack, could thwart our foreign policy objectives, degrade military performance, result in significant economic loss, and perhaps even undermine the confidence of our citizens in the Government's ability to protect its citizens and interests.

While no "smoking keyboard" has been found to validate such a threat, the very existence of the means to carry out such an attack, when coupled with the myriad of motives and opportunities that exist, results in our present state of vulnerability. These circumstances have created a situation that calls for prudent defensive actions to be taken in the public interest. We need to be proactive rather than be forced to react after an Information Age "Pearl Harbor." Moreover, a successful strategic attack would point the way and encourage others to plan similar attacks. Hence, we need to go on the offensive with a vigorous defense.

Digital War

Each age has seen war transformed by modern technologies and concepts. The Information Age promises to be no different. Some have called the Gulf War the first "Information War"—others have called it the last "Industrial Age" war. The power of information was clearly demonstrated in the context of traditional conflict. Information was leveraged to significantly improve the effectiveness of all aspects of warfare from logistics to command, control, communications, and computers, intelligence, surveillance, and reconnaissance (C4ISR).

The effectiveness of the U.S. and its allies in the Gulf War has surely deterred potential adversaries from taking on our forces in the rather symmetrical manner that Iraq attempted and has stimulated thinking about other strategies for countering conventional forces. Digital war, enabled by advances in technology and its widespread adoption as well as the globalization of economics and commerce, is surely a strategy that potential adversaries are thinking about to achieve some of the objectives that have previously been sought by means of traditional warfare.

Digital war, a subset of what we call information war, involves non-physical attacks on information, information processes, and information infrastructure that compromise, alter, damage, disrupt, or destroy information and/or delay, confuse, deceive, and disrupt information processing and decision making.

Digital war intrinsically possesses in ultimate form some of the same characteristics that traditional military planners are striving for, including low-cost precision, standoff, and stealth. Digital war threatens the ability of a nation state's military to interpose itself between its population and "enemies of the state," thereby causing a loss of sanctuary. The importance of sanctuary can be inferred by our willingness to spend significant resources on air, sea, and missile defenses to provide our citizens with a workable sanctuary with respect to territorial intrusions.

Another characteristic of information attacks stems from the loss of sanctuary. Attacks of this sort, particularly when they consist of more than an isolated incident, create a perception of vulnerability, loss of control, and loss of confidence in the ability of the state to provide protection. Thus, the impact can far exceed the actual damage that has occurred.

This non-linear relationship between actual damage and societal damage makes the problem of digital war a particularly challenging one because it creates a mismatch between rational defense responses and their effectiveness.

How does one respond to a serious set of information attacks? Responding with traditional military forces may be politically unacceptable or in fact, may be ineffectual. Currently there is no consensus, even among those in the defense establishment who think about these issues, regarding how to deal with such an attack.

Given the potential effectiveness of digital war, particularly as an instrument of power for niche competitors and non-state actors, we need, as a society, to take this Information Age form of war very seriously. If we do not, and if we rely solely on traditional weapons and concepts of war, we may be building our own 21st Century Maginot line that can be flanked with the speed of light.

Inadvertent Robustness

There are some who have suggested that we are not as vulnerable to information attacks as has been claimed because the collection of our legacy systems provides a certain amount of inherent robustness and resiliency. They point to the overlaps and duplications in these systems and argue that it would be very hard for anyone to completely disrupt a given set of services. They point to the lack of interoperability among legacy systems and the firewalls that are thereby created and argue that it would be impossible for attackers to get very far by penetrating the weakest systems and using them as launching pads for attacks on other systems. They argue that our current legacy systems and their interrelationships are difficult (even for us) to understand, so it must follow that potential adversaries will also be confused.

Clearly there is some truth in each of these arguments. But this unruly collection of legacy systems also carries with it significant disadvantages. As far as security considerations are concerned, five points need to be made. First, this issue is not whether or not an attack could totally destroy or disrupt a particular system or type of service, but whether or not there could be sufficient damage to trigger the perception of a failure and result in panic behavior that could in turn create a significant national problem. Second, the redundancies in the systems are only partial and unplanned. Hence, they are neither complete nor reliable. Third, our legacy systems, many having been designed and built with little or no attention to security, are difficult to protect and secure. Fourth, as the need for interconnectivity and interoperability increases, more and more systems are being lashed together with "work-arounds." These patches, in many cases, compromise security.

Lastly, the lack of security that these systems provide is dampening the demand for services that could make operations more effective and efficient in many areas. Our current collection of legacy systems has other disadvantages as well. For example, the lack of interoperability wastes resources and impairs operations. Thus, it should be clear that the disadvantages of our current collection of legacy systems are not a blessing in disguise but rather the source of problems that need to be addressed so that we can take full advantage of the opportunities that information technologies can provide.

Formulating the Problem

The first step in tackling any problem involves developing an understanding of the possible environments that may be faced (or the "states of nature"), one's options, and the objective that is being sought. This requires an identification of the variables that are relevant, that is, those that can significantly influence the outcome as well as the subset of these relevant variables that are controllable, which form the basis for designing options.

In a problem as complex as IW-D, working to formulate the problem accomplishes three things. First, it provides a useful framework for discussion. Second, it serves to keep the focus on those specific areas that are either unknown or in dispute. Third, it serves as a benchmark for measuring progress.

In this case, the *states of nature* correspond to the nature of the threat that will be faced vis-à-vis the vulnerabilities of our information infrastructure while our *options* correspond to the strategies we adopt and the actions we take to defend ourselves. The *objective* being sought corresponds to a level of infrastructure performance, its definition and measure being a major challenge in and of itself.

A good place to start is to try to develop an understanding of the nature of the threat, or more accurately the spectrum of relevant threats. This involves the identification of potential threats and the estimation of their likelihoods. Normally one would construct a set of states of natures that are mutually exclusive and collectively exhaustive so that a probability density function could be used. For the purposes of this discussion, the states of nature referred to correspond to potential threats grouped in some logical fashion to facilitate analysis of how well each defensive strategy deals with each of these threats.

Having an initial concept of the nature and range of potential threats, one can develop alternative defensive strategies and corresponding sets of action to counter one or more of these threats. A great deal depends upon what variables we believe we can and should control.

Each defensive strategy, with its corresponding set of actions, then needs to be analyzed with respect to each of the threats. The results of these analyses will be a characterization of the results or outcomes from pursuing each of the defensive strategies with respect to each of the threats. These outcomes, which are basically descriptions of results (e.g., number of penetrations and their consequences), then need to be translated into value measures that represent their impact. These costs and benefits provide a rational basis for determining an appropriate defensive strategy. Much will depend upon how we measure success.

Given the central role that the threat topology plays in problem formulation, we will now turn our attention to examining this topology.

Threat Topology

A graphical depiction of the threat topology is presented in Figure 2, The Threat Space. The irregular nature of the space is meant to indicate that its boundaries are not well defined. Nevertheless, one can group threats in order of the seriousness of their consequences (from left to right). The consequences associated with a failure to counter a specific attack range, on the one hand, from isolated and limited consequences to, on the other hand, consequences of catastrophic proportions.

In a series of pictures, Figures 3 through 6, different aspects of the threat topology are depicted. In Figure 3, the Threat Space is divided into three areas. On the left side of the space we can group the vast majority of the threats that occur everyday. These *Everyday* threats, while exacting a certain price, do not pose a threat to our national security. On the right hand side of the threat spectrum is a small area that represents those strategic threats having national security implications. The third area contains

Figure 2. **The Threat Space**

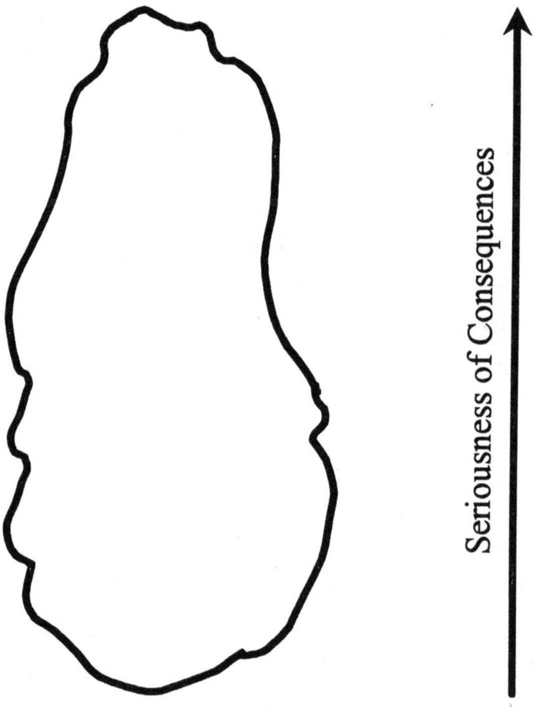

Figure 3. **Regions of the Threat Space**

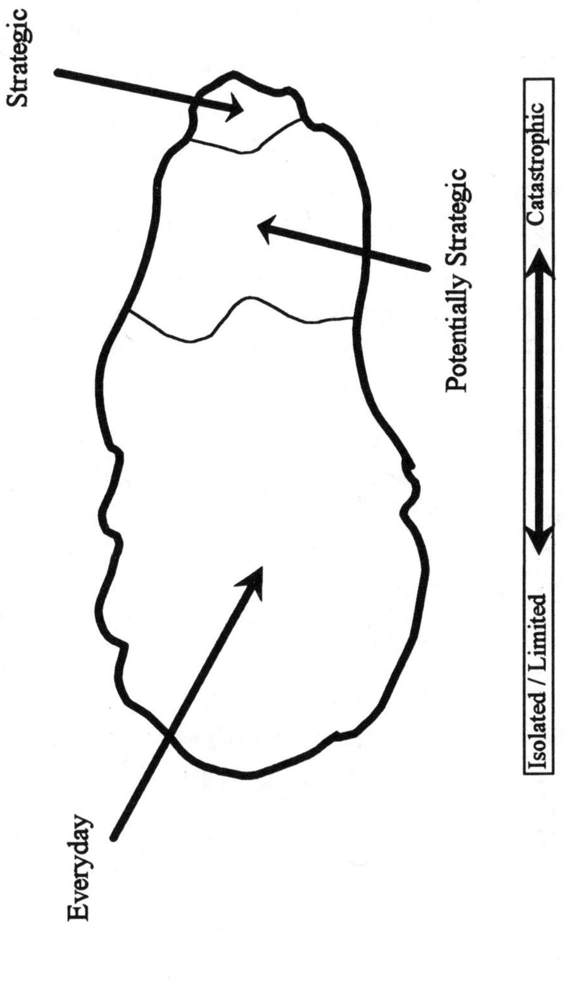

threats that may have national security implications. These *Potentially Strategic* threats represent a particularly difficult challenge.

There are a relatively small number of threats that most everyone would agree have strategic implications (Figure 4) and must be defended against with considerable vigor. Attacks against our systems that control and safeguard weapons of mass destruction (WMD) and our minimum essential emergency communication network (MEECN) clearly fall into this category. Others that fall into this category would include the information and communication systems associated with the National Command Authority (NCA), some of our command, control, communications and intelligence (C3I) systems, and some of our intelligence systems, particularly information regarding sources and methods. A review of other Government information and information systems would result in additional information and systems that should be added to this list.

Beyond those sets of threats that clearly fall into either the *Everyday* or *Strategic* categories, there are classes of threats that span the threat spectrum.

Attacks on our national, or for that matter international, infrastructure (Figure 5) do not fall neatly into one area of the threat topology but in fact populate all three classes of threat. These attacks on our public safety, energy, financial, and communications systems and services have different implications and

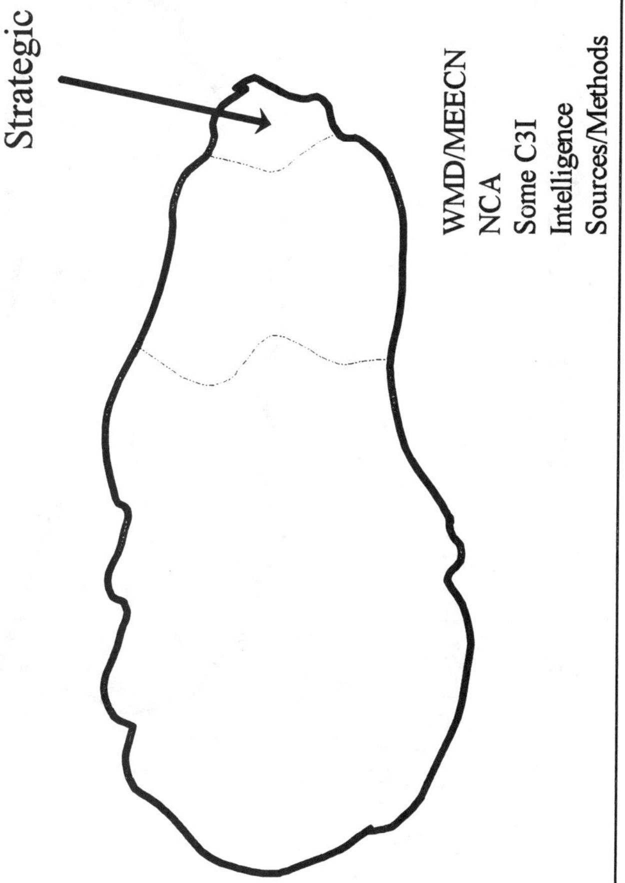

Figure 4. **Strategic Threats**

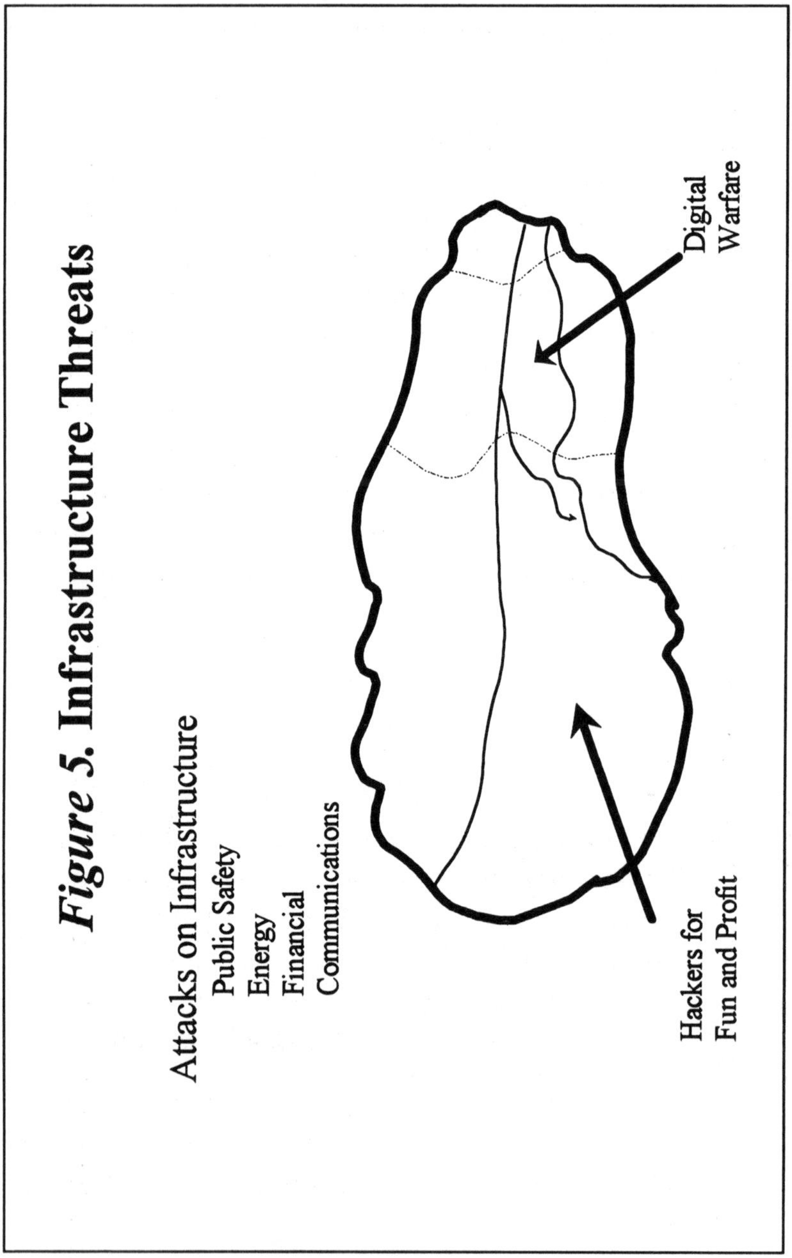

Figure 5. Infrastructure Threats

consequences depending on the specific nature of the attacks and the circumstances surrounding the attack.

The vast majority of attacks on infrastructure are by hackers whose motives run the full gamut from financial motives, to having some fun, or to more serious forms of antisocial behavior. While some of these attacks may have serious consequences in the form of significant losses of data, interrupted services, or stolen assets or services, only a small number of these lone perpetrator attacks are likely to have potential strategic consequences. This is not to say that it is impossible that some set of circumstances would result in the snowballing of one of these hacker attacks into a national security concern, but rather that this outcome is unlikely.

However, infrastructure attacks can be quite serious if they are well planned and coordinated. Arguably this would require an adversary with seriousness of purpose and with some sophistication and organization. This kind of attack would be better named digital warfare rather than be included as part of the group referred to as hacker attacks. Depending on the level of sophistication of a digital warfare operation, its consequences could range from a "high-end" hacker attack to an attack with strategic consequences.

A key point to be made involves the chaotic nature of the transition between topological boundaries for infrastructure attacks. Chaotic behavior involves a non-linear relationship between input and output where prediction becomes extremely difficult if not impossible. Two distinct scenarios serve to illustrate the chaotic nature of infrastructure attacks. In the first case, a particular infrastructure attack may trigger a series of proximate consequences that are difficult to predict and that greatly magnify the effects of the attack. In the second case, a series of attacks will exhibit chaotic behavior when the sum of their consequences can not be determined by adding up their individual consequences, or when their cumulative effect far exceeds the sum of the individual effects of a series of independent events. These are not uncommon patterns. Valid scientific disciplines of complexity, catastrophe, and complexity theory have been developed because these patterns occur throughout nature.

It is much the same story for attacks on commercial targets, depicted in Figure 6. In the *Everyday* category are attacks that amount to Information Age versions of fraud and theft—a continuation of white collar crime and a transformation of some more violent crime into a non-physical form. As Dr. Horton (PDASD (C3I)) has pointed out, one of the more notorious bank robbers of the 20th century, when asked why he robbed banks (given that they were often so well defended), was purported to remark, "that's where the money is." Well, digital money (assets and services) is where the money is in the Information Age.

Figure 6. **Threats Against Commercial Targets**

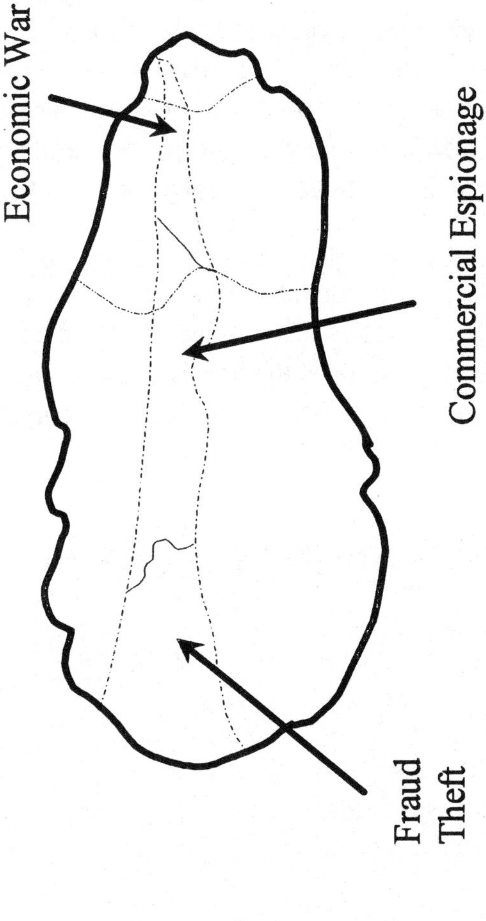

As with attacks on infrastructure, attacks on commercial targets can range from *Everyday* threats to *Strategic* ones depending on the circumstances. Attacks on commercial targets by competing organizations usually do not target money directly, but rather target vital information (e.g., trade secrets) and have the potential for more serious consequences than isolated thefts or embezzlements. These attacks, in the form of commercial espionage, have *Potentially Strategic* consequences, particularly when key industries are targeted by foreign companies.

Commercial espionage turns into economic war which could have *Strategic* consequences when it involves concerted efforts by state actors, international organizations, or other foreign-controlled groups.

Threat Characteristics

As can been seen from the series of graphical depictions of the threat topology presented above, the threat we face is multidimensional, somewhat messy and, with respect to the consequences of information attacks, can behave in a chaotic manner. The dynamic and interactive nature of the threat makes defending against attacks all the more challenging.

Attackers and defenders are locked in an ongoing battle of wits and resources as depicted in Figure 7, Threat Dynamics. Unfortunately, the attackers possess some inherent advantages. For example, the attacker can pick the time, place, medium, and method of the attack. The technology edge also goes to the attacker, for it is very difficult to develop perfect defenses at an affordable cost. Therefore, there will always be "holes." Which ones will be exploited are unknown until attacks occur, thus the

***Figure 7.* Threat Dynamics**

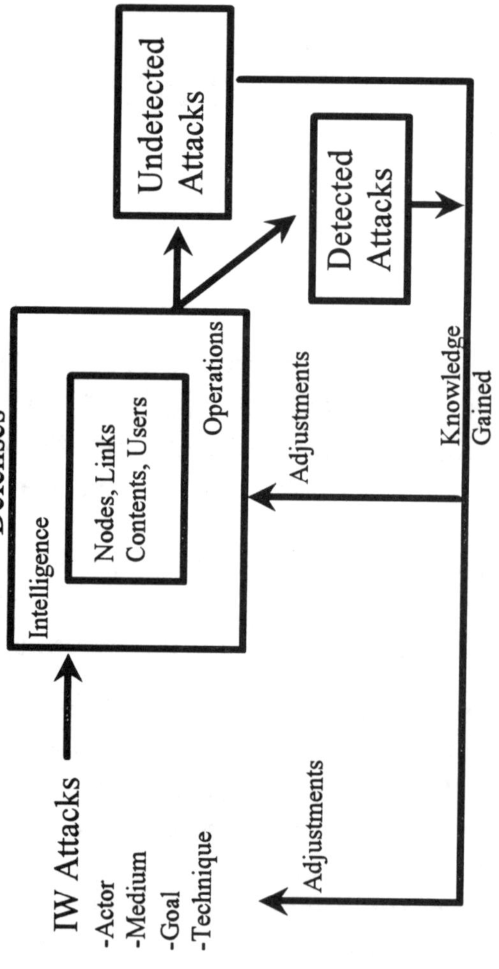

offense usually is one step ahead of the defense. Those who choose to orchestrate coordinated attacks on infrastructure also have the advantage that comes from being able to control their attack more easily than can a number of loosely coupled defenders.

In any event this is a learning environment for both attackers and defenders—a dynamic one at that. In this organic environment, attackers learn from undetected attacks, whether successful or not, while both sides learn from detected attacks, whether successful or not. Both attackers and defenders make adjustments and the "game" continues.

This aspect of the threat means that defense is not a one time thing. It must be a continuous activity. It also means that collection and analysis of information about attacks is vital to maintaining parity with attackers. Finally, it means that defenders must be proactive and undertake efforts designed to anticipate methods of attack so that timely defenses can be developed.

IW-D Considerations

The problem of how best to defend against a variety of potential digital information attacks is exacerbated by the following three characteristics of the problem: our inability to develop a simple relationship between a type of attack or threat and an organization, the essentially non-linear relationship between outcome and value, and the fact that key variables are not within our control.

Being able to decompose a problem often helps us to make progress, at least on some fronts, by facilitating the delegation of either functional or jurisdictional responsibility along organizational lines. As we have seen, the problem of IW-D resists decomposition along organizational lines. This is because there is a weak mapping between threat and organization from two perspectives. First, the organizational target of the attack is weakly related to the threat topology, that is, an attack on a given organization may result in a set of consequences that span the threat spectrum. Second, there is no clean mapping between organizational responsibility and the threat topology. This is particularly true of the area identified as *Potentially Strategic*.

It is always easier to solve problems with well-behaved or predictable objective functions or measures of value. As we have seen, the IW-D problem has situations where the relationship between events and variables behave in a chaotic manner. This introduces a large dose of uncertainty into the equation.

If these first two characteristics of the IW-D problem did not present significant challenges in and of themselves, we are also faced with the fact that some of the key variables that have an influence on outcomes and the values of these outcomes are only partially controllable. For example, each of the following variables, if fully controllable, could significantly either reduce the number or severity of attacks and/or the impact of the attacks: proliferation of technology, level of awareness and training of personnel, availability of computer expertise, system defenses, and public perceptions.

We have come to the end of the discussion of the nature of the problem and its characteristics that will drive the search for a solution. Next, a proposed solution approach is presented and discussed.

Proposed IW-D Strategy

The proposed solution to IW suggested here is a "Defense-in-Depth" strategy, a strategy that involves a series of successively stronger or "higher" defensive barriers that work together to decompose the spectrum of threats into manageable pieces. While implementing this concept still involves considerable challenges, it provides some structure for efforts to defend ourselves against information attacks.

Following a discussion of this IW-D strategy, the nature of a division of responsibility, some of the critical prerequisites for progress, key challenges that lie ahead, and the elements of an action plan for organizations with IW-D responsibilities will be addressed.

Defense-in-Depth

The proposed "defense-in-depth" strategy, depicted in Figure 8, consists conceptually of three lines of defense. Each line of defense is designed specifically to counter the threats associated with a particular region of the threat topology.

The first line of defense is to defend against *Everyday* attacks, which constituted most of the threat topology. Based upon the information available, the vast majority of these attacks can be handled with basic defenses.

The higher hurdles associated with the *Potentially Strategic* and *Strategic* attacks are then responsible for handling more sophisticated but far fewer attacks from fewer potential sources. For example, attacks with strategic implications would need to get through the first two lines of defense that should filter out all but the most skilled, resourced, and persistent adversaries. This means we can concentrate our intelligence and monitoring efforts on a smaller population, which in turn increases the chances of successful defense.

This defensive strategy also means that we can take different philosophical approaches with each line of defense depending on the nature of the threat. The two endpoints of the philosophical spectrum can be thought of as the "information first" and "security first" approaches. In the *Everyday* region of the threat

Figure 8. Defense-in-Depth Strategy

- Defense-in-Depth Approach
- Majority of Attacks Can Be Handled With Basic Defenses
- Higher Hurdles Handle More Sophisticated But Fewer Attacks From Fewer Potential Sources
- Mix of "Information First" and "Security First" Philosophies

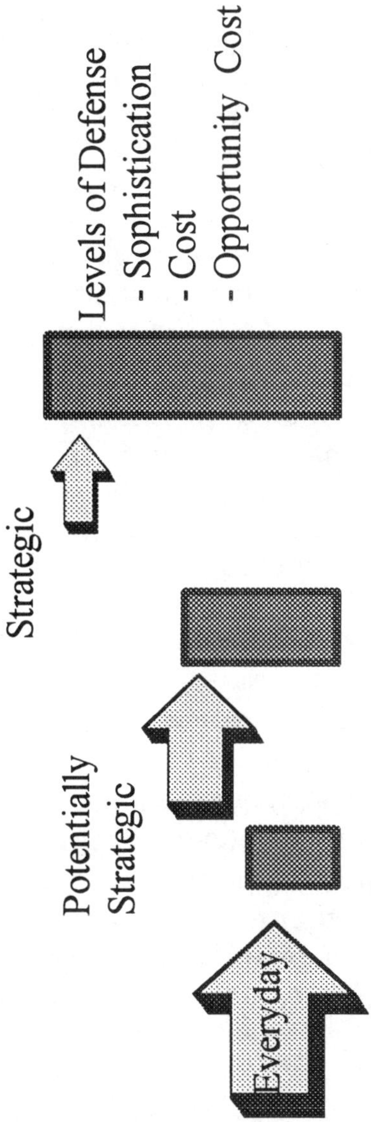

topology our approach has been to emphasize access to information. In the *Strategic* region, we put security first by restricting access and connectivity to the point of degrading performance and efficiency.

Division of Responsibility

Figure 9 graphically depicts a suggested division of primary responsibility for IW-D between the public and private sectors as a function of the threat topology. The modifier "primary" is used to make the point that, despite the assignment of responsibility in a particular area to either the public or private sector, both public and private organizations have responsibilities in each area.

The topological regions associated with either *Everyday* or *Strategic* threats are the most straightforward. Primary responsibility for the everyday threat should be the responsibility of the private sector. Handling such threats is simply the cost of doing business in the Information Age. With the availability of relatively low-cost defenses against these threats, the burden placed on the private sector is affordable. Furthermore, organizations are clearly in the best position to understand their own systems and the needs and concerns of their customers.

Responding to strategic threats is clearly the job of the public sector, although an adequate defense will involve some coordination with private sector and

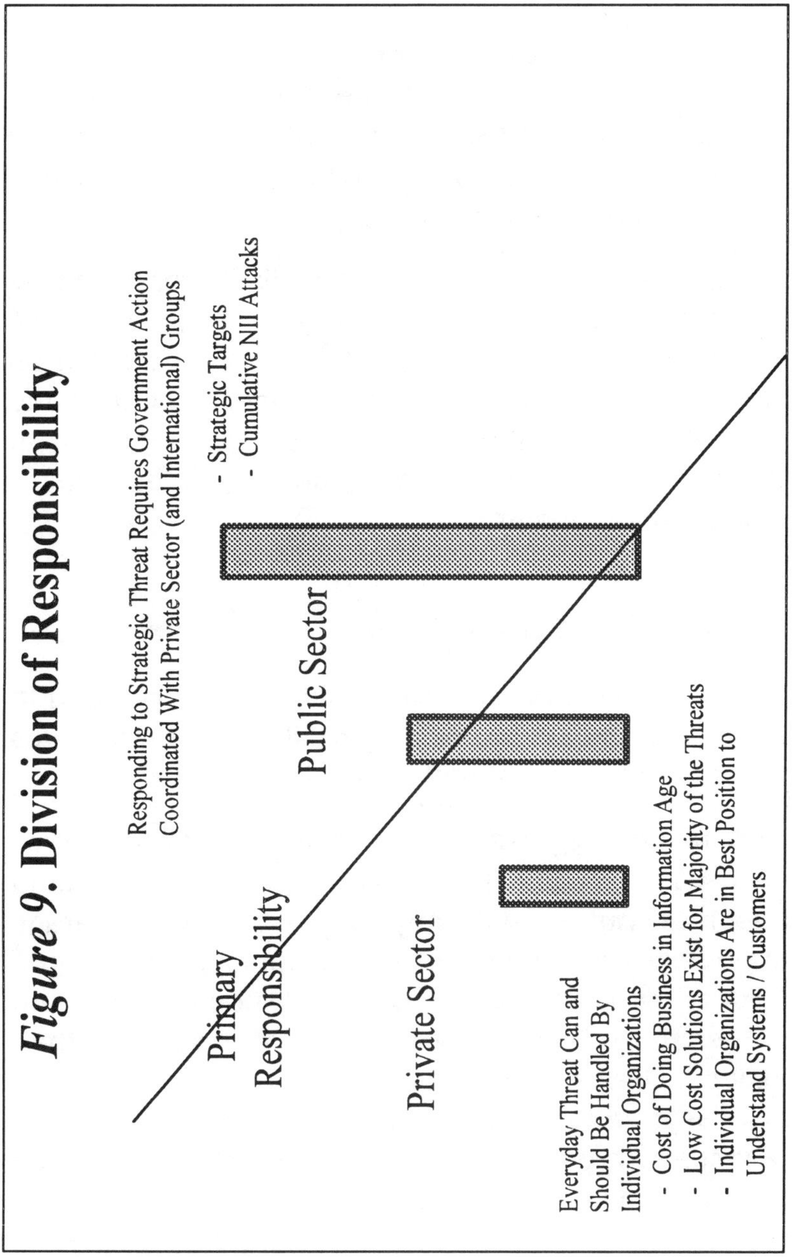

***Figure 9*. Division of Responsibility**

international organizations, particularly when it comes to the region of the threat topology that contains threats associated with attacks on the national information infrastructure or other institutions providing vital services.

Where to assign primary responsibility for defenses against threats in the *Potentially Strategic* region of the threat topology is less clear. This area could be called a "zone of collaboration," where the public and private sectors need to work closely together to understand the threat and develop mechanisms designed to counter it.

Perspectives on Information Security

Efforts to achieve effective collaboration will need to overcome the understandably different perspectives that organizations bring to the table.

Commercial organizations traditionally treat events such as *Everyday* attacks as simply a cost of doing business not significantly different than pilferage euphemistically referred to as inventory "shrinkage." Countermeasures have a definite expected value and are employed when their costs are less than their expected value. Private sector organizations traditionally respond to relatively low probability events with potentially large costs by either purchasing insurance or providing self-insurance.

The perspective on information security taken by organizations entrusted with information and systems deemed vital to our national security is quite different. Unlike many information attacks on private sector systems, the cost of a breech in security of national security information can not easily be determined. Overshadowing the actual costs of a particular incident is the fact that the very protection of the integrity of national security information and the systems that handle it is considered to have intrinsic value in and of itself. Risk avoidance is the ingrained response in these situations.

Given the nature of the IW threat topology, national and private sector information security are now inexorably intertwined. Attacks on the national or global information infrastructure can seriously affect private sector organizations and attacks on key private sector organizations that provide vital services have definite national security implications. These situations are contained in the *Potentially Strategic* region of the threat topology.

It is proposed that, in this region, rather than take a simple "dollars and cents" approach as in the *Everyday* region or a risk avoidance approach as taken in the *Strategic* region, we should take a collective risk management approach. This is clearly an area which requires defenses to be closely coordinated.

Managing the Solution

Given what we have seen about the nature of this problem as we have considered the different areas of the threat topology, it should be clear that any attempt to try to manage IW-D using a centralized approach is doomed to failure. If one has any doubt, they should be reminded of the weak threat to organizational mapping which prevents effective delegation of the problem with clean "chains of command" and the considerable limits on the ability of the Government or any single organization to control significant relevant variables.

Given the urgency and importance of this problem, a proactive stance is required. Therefore, it should be equally obvious that a laissez-faire approach is also doomed to failure.

A form of collective orchestration is needed to develop the degree of awareness and understanding of the threat and to develop the necessary defenses. The Government needs to lead by establishing what I have called a "framework for progress." The specific roles and responsibilities of the private and public sectors will vary as a function of the threat as we have seen in our brief look at each of the three major regions of the threat topology. While this discussion has focused on the roles, responsibilities and need for collaboration among U.S. players, the problem of IW-D transcends national boundaries. Without appropriate international agreements and cooperation among nations and international organizations, our collective ability to handle threats will be severely hampered.

Framework for Progress

While we have come a considerable distance in our journey to better understand the nature of this problem, many of us have been frustrated by the lack of a supportive environment for progress. Although we can continue to make progress, even on the rocky path we are currently forced to travel, progress in the six areas identified in the graphic will greatly smooth out our path and accelerate our progress.

First, one of the key prerequisites for progress is to create awareness of the problem and its complexities, as well as to foster a climate that will facilitate discussion and cooperation among the many groups and organizations that need to be a part of this effort. Given recent events surrounding some aspects of information security, we need to start by rebuilding bridges between some public and private sector groups and organizations.

Second, it is important that we work toward a well-defined vision that clearly lays out what we are trying to achieve and the appropriate role of the Government.

Third, the "rules of the game" need to be developed and promulgated. Many of our current laws and regulations have not caught up with the realities of the Information Age. A set of rules needs to address the establishment of information security standards, or a minimum level of defense to be associated with different kinds of data and information services. These would be similar to the recent development of privacy standards.

Fourth, self-interest, even enlightened self-interest and the desire of individuals and organizations to be good citizens, are not enough to ensure that appropriate actions and defenses will be developed and employed. Resources need to be provided for government organizations to help implement this framework for progress and to develop and implement the needed defenses. We also need to provide incentives that encourage public sector organizations to do what is collectively needed. In some specific cases, the Government will need to actually provide funds to private sector organizations to implement enhanced security.

Fifth, the solution to this problem depends on a great deal of cooperation among disparate groups and organizations. Mechanisms to facilitate and enhance cooperation (including the establishment of panels, groups, and clearinghouses) need to be developed.

Sixth, we need to fix responsibility for the many tasks involved in IW-D. We need to decide questions of jurisdiction. We need to make liabilities known and well defined. Finally, we need to clearly establish the responsibility of each organization. The nature of organizational responsibilities is discussed in more detail below.

None of these six aspects of the framework for progress is likely to be accomplished anytime soon. One only need review the legislative process and experiences with the translation of privacy concerns into a set of rules of the game to realize that it will be quite a while before each of these foundational pillars is in place. However, we must begin now to foster discussion of these issues and try to keep attention focused on this subject.

Allocation of Responsibilities

The responsibilities of public and private sector organizations differ as a function of the threat region. In all regions of the threat from everyday to strategic, however, each sector of society has some responsibility.

Responsibilities for Everyday Threats

The primary responsibility for the *Everyday* region of the threat topology falls upon the private sector. First and foremost, private sector organizations must assume responsibility for the protection of their own systems. When security laws and regulations are legislated and formulated, these organizations will, of course, also be responsible for adhering to these rules of the game.

Given the time it may take to develop and put in place a legal and regulatory framework to deal with the myriad of information security issues, it is proposed that on a voluntary basis, private sector organizations assume the responsibility for reporting incidents. It is hard to overstate the importance of the collection of information related to information attacks and their analysis. Without the development of a body of knowledge concerning these attacks, efforts at building defenses will be severely hampered.

The Government (including the federal, state, and local levels) must assume certain responsibility for this region of the threat topology as well. Clearly, the Government bears the responsibility for protecting its own systems and for the enforcement of appropriate laws and regulations. Given the importance of gaining international cooperation on this problem that knows no state boundaries, the Government must take on the negotiation of the necessary treaties and agreements.

Clearly, the collection of incident data with respect to its own systems is also a Government responsibility. But given the importance of pooling information to gain a more accurate situation assessment, the Government must also put in place appropriate mechanisms for data sharing, analysis of data, and the dissemination of results. Issues related to classification and security of these data and the products of these analyses will need to be addressed. A way must be found to get the information that individuals and organizations need to defend themselves to those involved in the effort.

Responsibilities for Strategic Threats

The Federal Government has the responsibility for the defense of the *Strategic* region of the threat topology, albeit with some support in selected areas from the private sector and state and local governments. Given the dynamic and interactive nature of this situation, it is important that the current and emerging threat be as fully understood as possible. Therefore, utilizing information collected as well as information reported by others, the Federal Government has the responsibility to perform strategic threat analyses on an ongoing basis.

The Federal Government also needs to develop an appropriate deterrence strategy designed to dissuade potential attackers. Strategic systems must be monitored and surveillance operations must be mounted.

Obviously, the Federal Government has the responsibility for protecting strategic information and the systems that collect, store, process, and disseminate this information.

Finally, the Federal Government needs to develop plans for reconstitution of damaged or disrupted systems and lost, compromised, or damaged information and for implementation of these plans should an event occur.

The private sector also has a role to play in this region of the threat topology. First, many strategic systems depend to some extent on the availability and integrity of private sector or state and local government information and systems. Second, some private sector or state and local government information and systems may be so critical that they are, for all intents and purposes, strategic. In both these cases, organizations need to cooperate with the Federal Government to protect these systems and the information they handle. Developing an adequate understanding of the threat requires that all organizations report incidents and share data related to attacks, whether successful or not. Thus another responsibility that must be assumed by private sector organizations is the prompt reporting of incidents and related information.

Responsibilities for Potentially Strategic Threats

The division of responsibilities for this region of the threat topology is not as clear-cut or as familiar as for the other two regions. In this zone of collaboration, public and private sector organizations need to find ways and create mechanisms that foster a shared perception of the nature of the threat, particularly those aspects of a situation that increase the likelihood of strategic consequences. In addition, ways should be found to enhance our defenses and improve our ability to mitigate the effects of attacks in this region to prevent them from having strategic consequences.

In this region, the Government needs to take the lead in helping to develop the necessary understanding of the threat, while the private sector needs to support Government efforts by providing incident data. The Government also needs to take the lead in developing coordinating mechanisms designed to both support improved understanding and the coordination of defenses.

The two biggest challenges associated with this pivotal threat region are first, the recognition of when an attack of strategic significance has begun and in characterizing the nature and scope of the attack and second, in the effecting of a transition from a "peace time" to a "war time" footing. The success of the processes and mechanisms developed to coordinate this transition will be critical to the success of any IW-D strategy.

The private sector needs to take the lead in this region by turning improvements in understanding into more effective defenses. This includes not only enhanced detection tools and techniques but also an improved ability to contain an attack, thereby limiting its spread, damage, and consequences. The Government needs to assist the private sector in these efforts by providing resources and technical support. Resources might take the form of tax incentives, or as was recommended in the case of some private sector systems that were deemed to be strategic, direct payments.

IW-D Challenges

For anyone who thrives on challenges, defending against information warfare is a great line of work. There are five key challenges we face. The first challenge involves the development of a better understanding of the nature and characteristics of the threat among society and its institutions. Success requires that everyone be on board. Therefore, it is important that we continue to work to increase awareness of this problem and to develop a better understanding of both the nature of the threat and our vulnerabilities.

The second challenge is to develop a strategy for deterring digital information attacks. The first line of defense is deterrence. Not enough effort is being devoted to developing and gaming possible strategies. In February 1996, ACTIS sponsored a workshop on this subject. This workshop was a highly successful endeavor, giving those who attended a better idea where the latest thinking is on this subject, stimulating more thinking about the subject, and bringing some key issues into sharper focus. (The proceedings of this workshop, "IW and Deterrence," have been published and are available from ACTIS.)

The third and fourth challenges involve developing means of providing warning for attacks and ways of successfully defending against attacks that do occur. Improving our ability to see an attack coming, or providing indications and warning (I&W) of attacks in a timely fashion, is perhaps the most single difficult challenge we face. Developing an I&W capability involves not only the traditional strategic and tactical warning capabilities, but also the ability to know that an attack has begun and to ascertain the likely scope of such an attack. Given that currently, in many cases, an attack in progress is not even recognized, this will be a tall order. This is not only a technical challenge, but it is, as mentioned earlier, an organizational challenge. This is because of the information necessary to provide warning of an imminent action or an attack in its early stages would most likely need to come primarily from private sector organizations whose "peace time" reporting structures may not either require reporting of all relevant incident data or require it in near real time. Thus an attack of strategic significance may be well underway before we realize it and are able to move to a "war time" footing, which would bring with it the increased reporting requirements and coordination necessary to assess the situation and respond accordingly.

As a result of my participation in a series of "day after" games developed by RAND for ASD(C3I), I have concluded that we will not be able to respond to such attacks in a timely and effective manner unless there is 1) more awareness and understanding of the nature of strategic IW capabilities and our own vulnerabilities among not only key officials in public and private organizations but also among Congress, the Media and the Public; 2) more understanding of the offense arsenal at our disposal, particularly the direct, indirect, and collateral damage they might cause; 3) a pre-agreed systematic system of alerts similar to the DEFCON system, each level of which carries with it known, understood, and practiced processes and actions; and 4) a "battle damage" assessment process suitable for IW.

The construction of a DEFCON-type system for IW-D will require the investment of a considerable amount of intellectual capital as well as a considerable amount of coordination among a wide variety of Government departments and agencies, the Congress, industrial associations, and private organizations. Given the nature of the problem, time will be of the essence should problems with our information infrastructure begin to appear. Well thought-out options that address both the need for increased collection and analyses and the need to take measures to prevent or control damage are essential to countering this threat. Given the trifurcated threat topology and the very different nature of each of the three threat regions, implementing the proposed "defense-in-depth" strategy will be a considerable undertaking.

The fifth challenge is to develop appropriate and effective responses to such attacks. Responses to attacks include identification, interdiction, apprehension, and punishment (possibly including retaliation).

IWS Awareness and Understanding

We have much to learn and many to educate. When many of the individuals who need to become more aware of the threat and its potential consequences are exposed to the subject only by reading novels or going to the movies, we cannot really expect to develop the degree of understanding required. When the only exposure to the subject is through fiction, it is no wonder that the threat may be dismissed as fictional. There are still many individuals in key positions in both the public and private sector who need to have a better appreciation for this problem and to be more motivated to address the issues.

On the other hand, admittedly we do not possess a great abundance of factual information. While we have clear indications that some potentially serious attacks, even crippling attacks, are technically feasible, as has been pointed out, there is no "smoking keyboard" to show. Yet it should be pointed out that the time it took to create a working atomic bomb from the time its theoretical feasibility was recognized surprised many, even the most knowledgeable scientists.

Our ignorance about the nature of potential attacks is mirrored by a lack of knowledge about the effectiveness of current and developing defensive techniques and strategies.

When our systems are not being adequately monitored and incidents are not being adequately recorded and investigated, it is hard to see how we can develop the vastly improved understanding of both the threat and the effectiveness of defenses we require. Increased collection and analysis is clearly needed to provide the empirical foundation required to a) increase awareness, b) increase our understanding, c) support planning, and d) develop effective defenses. Given our tendencies to value privacy, both in terms of individuals and corporations, and the limitations we have imposed upon the collection of domestic intelligence, it is unlikely that sufficient information will be collected and analyzed by the Government to quickly determine the nature and extent of an ongoing IW attack in its early stages. Therefore, it is extremely important that when it appears that such an attack may be imminent or in progress, that "peace time" data collection and analysis arrangements be quickly shifted to a "war time" footing. In designing a DEFCON-like process, it will be important to define "trigger" events that will automatically cause information to flow to a situation assessment organization. This is because of the time criticality of providing up-to-date information to the decision makers who will be considering appropriate DEFCON-like decisions involving potentially costly actions to safeguard the NII and prevent further damage to our information infrastructure and the key public safety and economic functions that it supports.

IW Deterrence

With the dawn of the atomic age came the recognition that developing strategies for deterrence and counter proliferation needed to be pursued with a sense of the utmost urgency. IW differs from atomic warfare in a number of significant ways, and therefore lessons learned from our experience in developing a workable strategy for deterrence may not apply directly to the problem of deterrence of IW attacks, but certainly may provide a starting point or checklist for consideration.

Some of the compelling issues related to the development of a deterrent to IW attacks include various means of raising the attack threshold, using offensive IW, dealing with non-state actors, taking preemptive actions, and developing potential forms of punishing attackers.

While raising the defensive threshold, thereby making attacks more difficult and costly as well as limiting the damage they can do, is widely recognized as an important component of any deterrence strategy, an issue that needs to be addressed relates to the "height" of the threshold. For example, what is more defense? When does more defense become counterproductive?

Another critical issue is whether or not having and indicating a willingness to employ a potent offensive IW capability would be an effective deterrent, and if so, in which particular set(s) of circumstances.

Given the low cost and small footprint required, non-state and even individual actors may gain the wherewithal to pose a strategic threat. How can one gain the leverage on these kinds of adversaries to deter them from launching such attacks?

Other key issues include the nature of preemptive actions that could be employed and the relationship between punishment (or retaliation) and deterrence.

Building a Defense-in-Depth

Our ability to defend individual systems and the information they handle is the foundation of any IW-D strategy or approach. In order to achieve the requisite level of defenses, we need to 1) revise our approach to system design and acquisition, 2) understand the dimensions of defense, and 3) support the development of critical enabling technologies.

System Design and Acquisition

The current trend in designing systems is to provide vastly increased access to increasing amounts of information at all levels of an organization and to provide for more external connectivity. We in government are joining the move to separate the flow of information from the management or command structure. Our new approach to information systems involves "reach back," that is, individuals are not "pushed" information but rather "pull" the information they require from distributed databases. In response to the sheer amount of information available, more and more sophisticated presentations (displays) are being developed. Driven by the reduced life cycle of information technology and shrinking budgets, we are becoming more and more reliant on commercial-off-the-shelf (COTS) equipment and software. These trends make it more and more difficult to ensure the integrity of information or recognize when it has been compromised.

If we are to build the level of defenses to IW attacks we need and provide users with the level of assurance required to give them confidence in their systems, we need some changes in our approach to system design and acquisition. For a number of reasons, not the last of which is security, we need to focus more on providing quality information, that is on moving up the "data–information–understanding––knowledge" continuum. In evaluating system designs and in setting standards for systems to meet, we need to augment the current set of measures of merit (MOMs) for information. In addition to attributes such as timeliness and accuracy, we also need to focus on authenticity, integrity, and availability.

Connectivity is important to achieving the levels of functionality desired, but increased emphasis on secure interoperability is needed to achieve IW-D objectives. Fixing and patching systems and installing new releases offer many opportunities to bypass or disable system defenses. Operations and Maintenance (O&M) activities also provide opportunities to prevent intrusions and strengthen system defenses. System operators and software maintenance engineers are often the least experienced and lowest paid in their professions. They also typically receive the least training. Given the critical nature of O&M for maintaining system defenses, O&M practices need to be reviewed.

Dimensions of Defense

Defense is a function of more than design and software quality assurance. It has many dimensions, some of which are depicted in Figure 10. Each of these dimensions needs to be considered for each specific system and set of circumstances.

In addition to how a system is designed, these dimensions of defense include system operations, methods, and procedures employed to limit the attractiveness of an attack and/or the consequences of an attack. Figure 10 presents some of the dimensions of defense that need to be considered in constructing and "tuning" each system to its unique set of circumstances. Each dimension offers a range of choices that provides either more or less protection. More protection always comes at a price

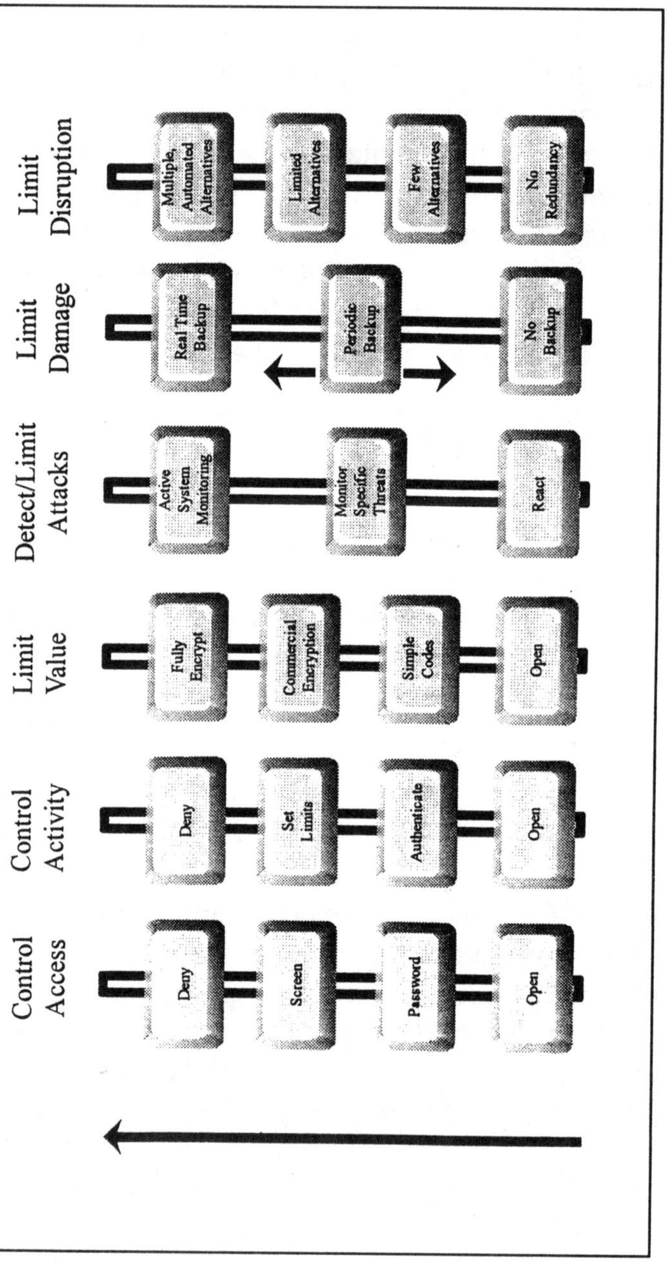

Figure 10. **Dimensions of Defense**

(although surprisingly the price may be quite low, particularly when compared with the security gains that can be achieved). Achieving higher levels of protection either actually costs more to build into a system or exacts a cost in terms of operating overhead or in loss of functionality.

Critical Technologies

Building defenses into systems presumes we have the means to do so. Many of the defensive capabilities we currently have are not adequate for certain known levels or types of attacks, not to mention technically feasible but undocumented attacks. The following are some areas in which we could use some advances in technology.

Real-time intrusion detection is clearly a key element in any set of defenses. Our ability to detect, in real time, intrusions into our systems and the identity of the intruder is currently very limited. In does not take very long to carry out an information attack. Damage can occur in an instant. Clearly an automated capability to respond to an intrusion that can prevent or limit the damage would be highly desirable.

Given our increasing reliance on COTS, we need ways to cost-effectively make sure that the software we buy does what we want it to and only what we want it to. Any Information Age organization buys millions of lines of code each year whose exact origins are not known with any degree of confidence. Automated tools for performing quality assurance (QA) and for verifying and validating (V&V) the code would be an immense help.

Knowing for sure that data were not altered or compromised and that the source of a piece of data or a message was verified would go a long way in the effort to combat certain types of IW attacks. More work needs to be done to provide cost-effective data and source authentication.

Organizational Action Plan

The cornerstone of our efforts to combat IW will be the efforts of all organizations to protect their own systems and information. Some organizations have been worrying about this for a long time and have developed and implemented plans to keep on top of this increasingly serious set of threats. Other organizations have more work to do.

It might be helpful, even for those organizations that feel they are well prepared, to review the following list of suggested actions to determine what they need to do to be better prepared for the future.

The first suggested action involves a review of the organization's mission in light of the emerging threat. A few organizations may find that IW-D adds a mission or increases the importance of an existing mission.

New relationships with external organizations may be required, or perhaps existing relationships may need to be modified. Thus, a review of these relationships is in order.

Who is responsible for IW-D in the organization? Perhaps the organization has a Chief Information Officer (CIO) and it would be appropriate for the CIO to take on this responsibility. Perhaps the responsibility for IW-D is spread out among several individuals. In any event, a clear allocation of responsibilities is required.

Not all information or all systems should be considered equal with respect to the protection they merit. It is important, given resource constraints, to identify which information and systems (and functions of these systems) are critical and which are not critical.

How vulnerable are the information and systems? What is the specific nature of the vulnerabilities? Answers are needed to provide a basis for planning and developing defenses. It needs to be remembered that vulnerabilities are relative to the threat, the nature of which is constantly evolving. Thus, vulnerability analyses are not a one-time task but must be part of a continuing effort.

Isolated actions to improve security are helpful, but they are no substitute for the development of a comprehensive IW-D strategy for an organization. Since it is not possible to avoid all the risks associated with IW, each organization needs to develop a plan to manage these risks. In the course of developing and articulating an organizational IW-D strategy and risk management plan, many issues will be raised and discussed. These discussions will create a greater awareness of the problem within the organization and improve the organization's ability to meet the challenges associated with IW-D.

Combatting IW is a long-term proposition. There are many long poles in the tent. An organization's investment strategies need to be reviewed and investments in defenses and supporting technologies must be made. Some reallocation of resources may be made necessary by changes in the operating costs associated with introducing new procedures and safeguards.

Summary

The problem of defending against information warfare is real. Our citizens and the organizations that provide them with the vital services they need can find no sanctuary from these attacks. The low cost of mounting these attacks has enlarged the field of potential adversaries and complicated efforts to collect intelligence and array our defenses. The consequences of a well-planned and coordinated attack by a relatively sophisticated foe could be serious. Even the threat of such an attack or digital blackmail is a distinct possibility. How the public will respond to the threat of IW infrastructure attacks or to actual attacks is unclear, but is a major determinant of future policy and actions.

This situation is getting worse with the rapid proliferation of information technology and know-how. We are becoming increasingly dependent on automation in every aspect of our lives. As information technology becomes an essential part of the way organizations and individuals create products and provide services, the need for interconnectivity and interoperability increases. With this increased need for exchanges of information (and products), vulnerabilities increase. Finally, the increased reliance on commercial-off-the-shelf products or commercial services makes it more and more difficult for organizations and individuals to control their own security environment.

Given this situation we need to focus on two goals. First, we need to find a way to protect ourselves against catastrophic events. Second, we need to build a firm foundation upon which we can make steady progress by continually raising the cost of mounting an attack and mitigating the expected damage.

About the Author

Dr. David S. Alberts is both Deputy Director of the Institute for National Strategic Studies and the Director of Advanced Concepts, Technologies, and Information Strategies (ACTIS) at the National Defense University, which includes responsibility for the School of Information Warfare and Strategy and the Center for Advanced Concepts and Technology. He also serves as the executive agent for the Department of Defense's Command and Control Research Program.

Dr. Alberts has extensive experience developing and introducing technology into private and public sector organizations. This extensive applied experience is augmented by a distinguished academic career in computer science and operations research and government service in senior policy and management positions.

*U.S. G.P.O.:1996-405-201:40010